SPACE DEBRIS

Environment / Biomimicry

Malgorzata Juliet Stach, kursist.nr. 20150517, gruppe nr. C1-113-5

AALBORG UNIVERSITET ESBJERG, HOLD AK 1årlig adgangkursus

COPYRIGHT

First published 01.05.2016

Written 23.03.2016 by Malgorzata Juliet Stach

Malgorzata J.K.J.J.G.Stach

ms@stach.dk

www.stach.dk

ISBN 978-1-365-07126-3

Typeset by Malgorzata J.K.J.G.Stach
Printed and bound in www.lulu.com
3101 Hillsborough St. Raleigh, NC 27607
919-447-3305
http://www.lulu.com

DEDICATION

To The European Astronaut Corps is a unit of the European Space Agency (ESA) for select, training, and providing astronauts as crew members on U.S. and Russian space missions.

Andreas Enevold Mogensen, Ph.D. a Danish Engineer and first Danish astronaut with AK-student of Aalborg University in Esbjerg Malgorzata Stach.(Foto 27.02.2016)

Contents

Contents

THE GRAVITY

Gravity is a very powerful force. Every object in space exerts a gravitational pull on every other, and so gravity influences the paths taken by everything traveling through space.

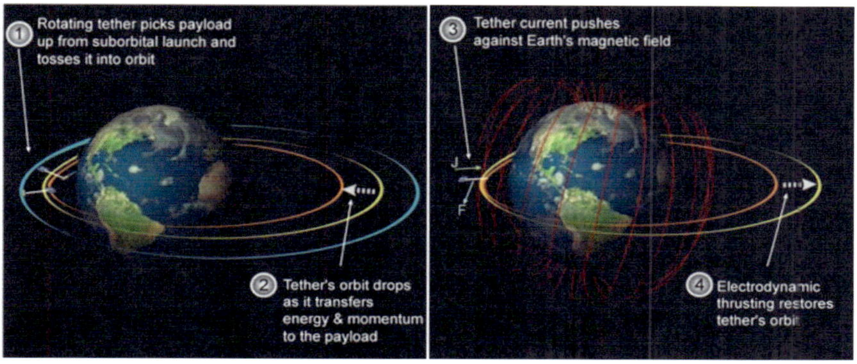

Figure 1

It is the glue that holds together the entire galaxy. It keeps planets in the orbits. It makes planets habitable to trap gasses and liquids in an atmosphere. It can also cause life-destroying asteroids to crash into planets. Gravity from the Sun reaches throughout the solar system and beyond, keeping the planets in their orbits. The reason that astronauts feel weightless,has actually to do with the gravity and distance to the earth. We feel the weight on Earth because gravity is pulling us down while the floor or ground stop us from flying or hanging in the air.

Figure 2

Scientists know the basic laws of physics and movement rules that objects follow in space. Gravity is the most important force affecting the movement in space. The strength of gravity's pull between two objects depends on how big they are, and how close they are.

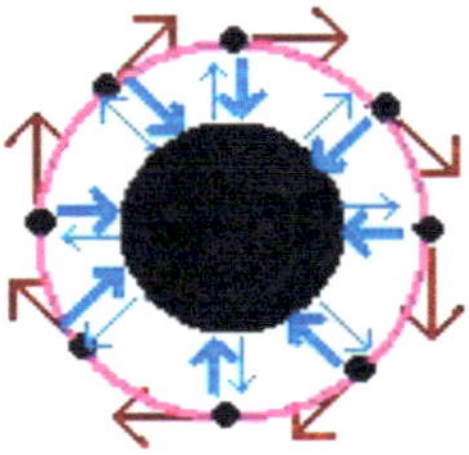

Figure 3

Figure 4

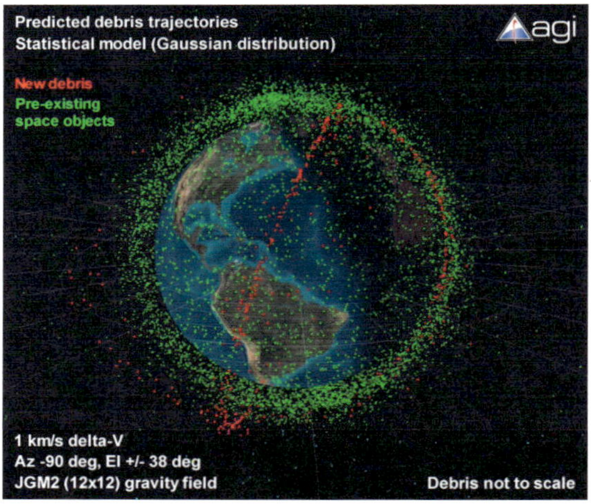

Figure 5

The space debris is flying around the Earth and stays within the planet´s gravity field. (Figure 3 and 5)

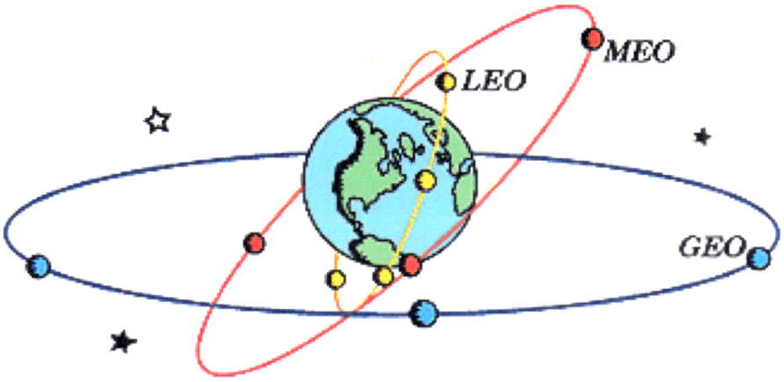

Figure 6

(Figure 6) The space around the earth is filled with more than 300.000 artificial debris particles with the diameter larger than 1 cm. The space, where the debris is placed, is divided into three main regions: the Low Earth Orbit (LEO), below about 2 000 km, the Medium Earth Orbit (MEO), between 2 000 km and about 36 000 km, and the Geosynchronous Earth Orbit (GEO) at about 36 000 km of altitude.

THE SPACE PROJECTS ON DEBRIS

In 1978, NASA scientist Donald J. Kessler proposed an ablation cascade or collisional cascading tat is called The Kessler Syndrome. The density of the objects in LEO is enough that collisions between objects could cause a cascade - to generate debris for further cascade by the use of satellites in specific orbital ranges. This is a long term project and it takes many generations. The Envisat satellite had been made to contribute debris from a collision during the next 150 years. By using the multi-megawatt land-based laser broom, it is possible to re-orbit the space debris. The laser would ablate and create a thrust that would change the eccentricity of the remains of the fragment until it would re-enter harmlessly In the year 2007, China constructed an anti-satellite missile test by shutting on space debris from the ground. (see figure 7)

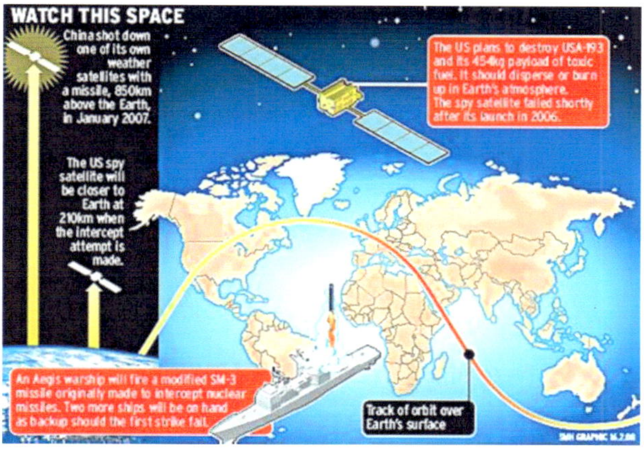

Figure 7

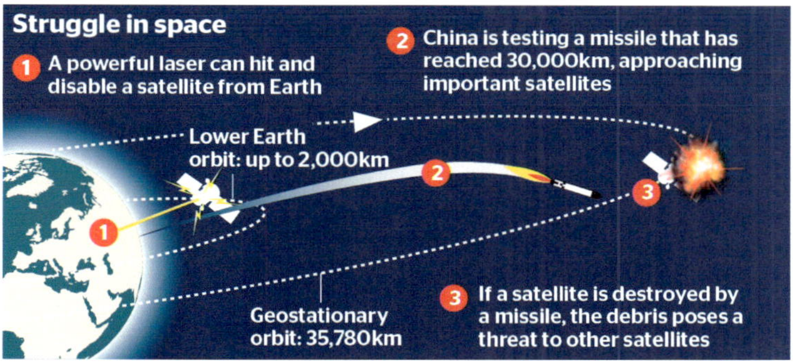

Figure 8

THE SOLUTIONS OF REMOVING SPACE DEBRIS FROM ORBITS

It is already possible to destroy the space debris by shutting missiles on the target from the military base on the Earth. This is not such expensive as sending a new satellite with robots. But it is a waste of the material, time and work with manufacturing.

Every project needs calculation and imagination. Space is huge, and even our immediate environment is gigantic and expensive. Experts write on the internet that there is no money to clean the Earth´s field of the debris. But in the longer term, there will be not enough money to repair satellites damaged by the debris. Actually, it is dangerous to work in the satellites, when space debris is flying around and damaging astronauts´ working place.

Figure 9

Figure 10

What if space debris is protecting the planet from star dust, small asteroids, radioactivity from space and sun spots? What if the space debris makes the earth´s gravity field stronger and protects against moon´s activity for tsunamis, earthquakes, rapid dissolution of the ice on the South Pole and the North Pole and rising oceans? What if scientists and the government do not want to clean earth field from space debris? What if the project is waiting for an unknown project planned in the future? What if it is dangerous for humanity to send the space debris back to the earth? The reason can be radioactivity of the debris. It can also be dangerous for humans being hit by metal parts falling down / decreasing in high speed to the Earth without human control over the precise destination.

The second possibility to get off with the space debris is to send it to another planet or the sun. This project is expensive and does not give the result for recycling or environment. In the year 2018, Nasa will send the probe to the sun´s 1400C atmosphere for exploring purpose. The project Solar Probe Plus will give many answers about the solar system.

Figure 11 - The Solar Probe Plus mission has slated to launch no later than 2018, and will visit and study the sun closer than ever before.

THE PROJECT - SUN

Earth is the third planet from the Sun and the third of three inner planets. The planets are far away from the Sun. Traveling takes a long time and needs physics calculations. Figure 12 and 13 shows the planets in their orbits on the orbital plane that is called ´´the inner solar system´´.

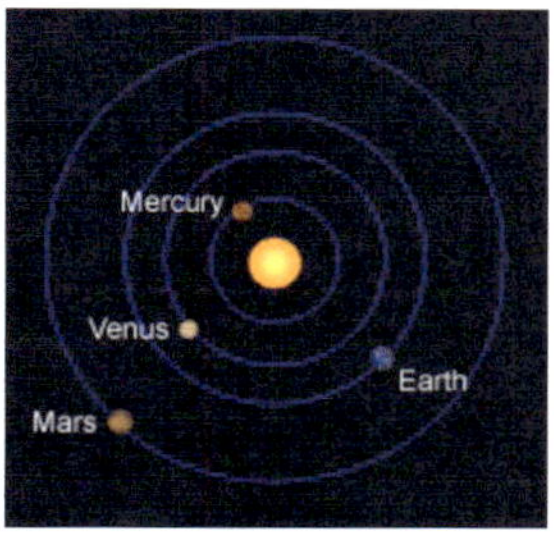

Figure 12

Figure 13

OBJECT	Distance from Sun (average)	Distance traveled in one complete orbit of the Sun (one "year.")	Amount of time for one complete orbit of the Sun (one "year.")
Sun	0 miles		
Mercury	36.800.000 miles	223.700.000 miles	3 Earth months
Venus	67.200.000 miles	422.500.000 miles	7 Earth months
Earth	93.000.000 miles	584.000.000 miles	1 Earth year (365.25 days)
Mars	141.600.000 miles	888.000.000 miles	23 Earth months. Almost 2 Earth years.
Jupiter	483.600.000 miles	3.037.000.000 miles	142 Earth months. Almost 12 Earth years.
Saturn	886.500.000 miles	5.565.900.000 miles	354 Earth months. (29.5 Earth years)
Uranus	1.783.700.000 miles	11.201.300.000 miles	1009 Earth months. (84 Earth years)
Neptune	2.795.200.000 miles	17.562.300.000 miles	1979 Earth months (almost 165 Earth years)
Pluto	3.670.100.000 miles	22.698.700.000 miles	2977 Earth months (248 Earth years)

The Sun had formed about 4.6 billion years ago from the collapse of a giant molecular cloud that consisted mostly of hydrogen and helium and that probably gave birth to many other stars. That means that our solar system is 4.6 billion years old. The sun is the center of our galaxy and is the hottest planet in our solar system; furthermore, the Sun contributes life on the Earth.

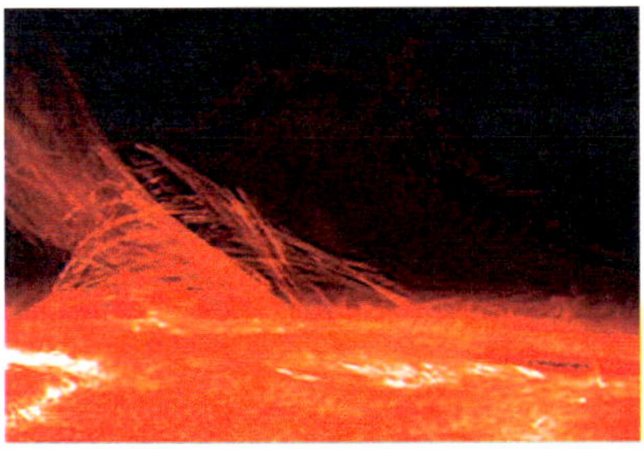

Figure 14

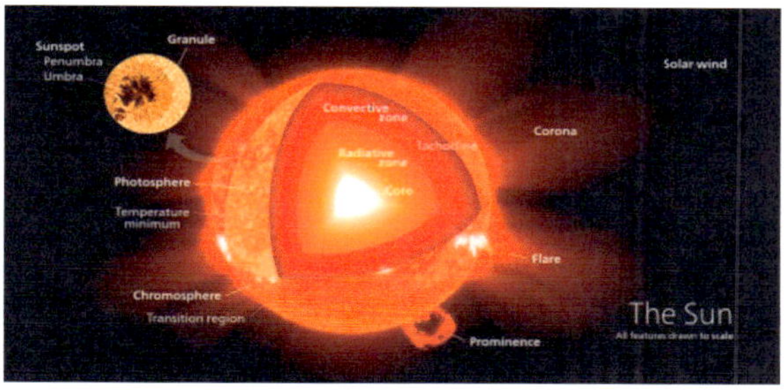

Figure 15

Nothing can survive getting very close to the Sun. It is impossible to explore the sun´s surface in person, because of the high temperature. (Figure 14) The sun is closest like the lava in volcanos. The distance from The Earth to the sun is 93.000.000 miles.

The core of the Sun (Figure 6) extends from the center to about 20–25% of the solar radius. It has a density of up to 150 g/cm3 (about 150 times the density of water) and a temperature of close to 15.7 million kelvin (K). By contrast, the Sun's surface temperature is 5,800 K. The Sun has a magnetic field that varies across its surface. Its polar field is 1–2 gauss (0.0001–0.0002 T), whereas the field is typically 3,000 gauss (0.3 T) in features on the Sun called sunspots and 10–100 gauss (0.001–0.01 T) in solar prominences. The solar magnetic field extends beyond the Sun itself. The electrically conducting solar wind plasma carries the Sun's magnetic field into space, forming the interplanetary magnetic field. That is holding the whole solar system in balance. The Mariner 10 spacecraft was built to use this method, using Venus' gravitational pull to slingshot it towards Mercury in February of 1974. In the 1980s, the Voyager 1 probe used Saturn and Jupiter for gravitational slingshots to attain its current velocity of 60,000 km/h (38,000 miles/h) and keeps it into interstellar space. The distance from the Earth to the sun is 93.000.000 miles. There is no gravity in outer space, therefore, it is possible to move spacecraft faster than 38,000 miles/h.

THE POSSIBLE SOLUTION OF THE NEW SATELLITE

Scientists had found many methods to localize, recognize and observe space debris.

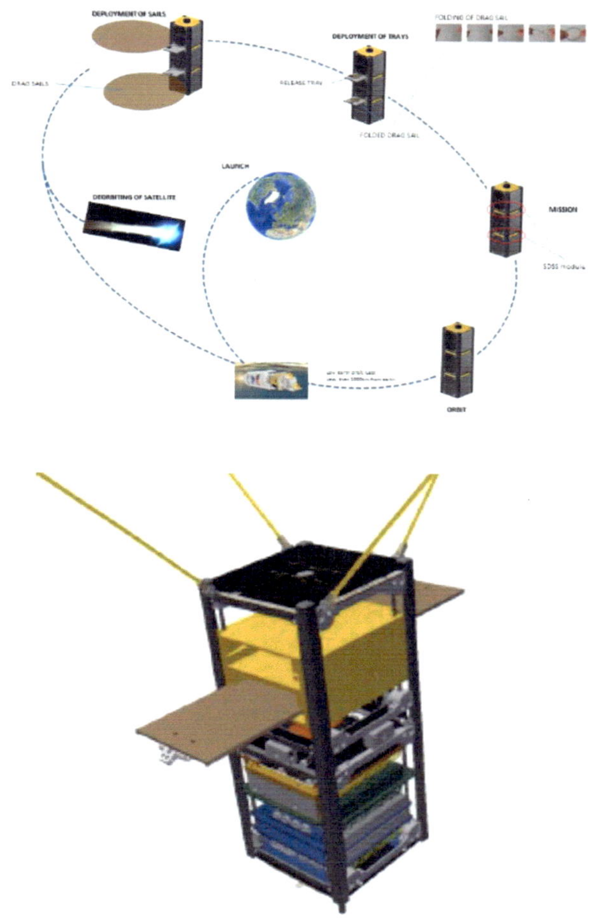

Figure 16 - Overall satellite mission from launch to deorbit. The drag sail principle is illustrated for a CubeSat satellite configuration (The NanoRacks-GOMX-2 spacecraft fitted with 2 SDSS modules placed in-line (depicted without sails).

The collection of space debris of size more than 1cm requires a magnet or a machine with the help of human- and robotic Astronauts and a satellite or ship to transport the debris to the sun or to the earth for recycling. The Japanese technology allows robotic astronauts on the satellites. They help to repair satellites inside and outside in space. First robots were very expensive, but after a few years, the prices fell down to the minimum possible. People use robots to the private purpose and robots in fabrics are the ´´must´´.

By building a satellite controlled by robots, can be possible to assemble/collect the space debris. Figure 17-18 shows the satellite, which is controlled by the internet and powered by solar panels. The satellite is made of a metal, has a magnet and adhesive textile build in. The adhesive textile will catch metals that are resistant to magnetism. The ship operated by robots can be dropped down to the Earth as balls. The balls can also be sent to the sun.

By using biomimicry it is possible to create a robot that can be controlled by space internet, solar panels and human astronauts, later robot astronauts. The big robot looks like a spider and has a control panel for the ´´driver/operator´´, and a rocket motor as the power on the way to the sun. The small spider robots can function as the satellites and sensors for finding the debris (Figure 17-18). On the sides of the robot, there are magnets, adhesive textile or some kind of the spider webs. For junk larger than a few meters can the spider uses own parts/legs to keep the junk until it merges with the rest. (Figure 20). The spider vehicles are also the transporters of the space debris to the sun. That means that both debris ´´packages´´ in the big balls or in one big ball, operated by robots, will be melted to nothing when flying into the sun´s gravity field.

Figure 17

Figure 18

Figure 19

The satellite (Figure 20) is using solar energy and can be built with a little working place for the human astronaut or robotic astronaut. ESA is using space internet to control robots and the satellite.

Figure 20

The balls are released from the satellite and pushed out into space. Each ball is equipped with a magnet and a sticky plasma in the core and is rotating very slowly around its own axis. (Figure 21) The space debris is caught in the ball. The small replicator robots are sitting on the ball and are controlled by space internet to move debris inside the ball. (Figure 21)

When the ball is filled up, the openings close and the ball is redirected to the Earth´s coordinates by small satellites or the vehicles to SUN.

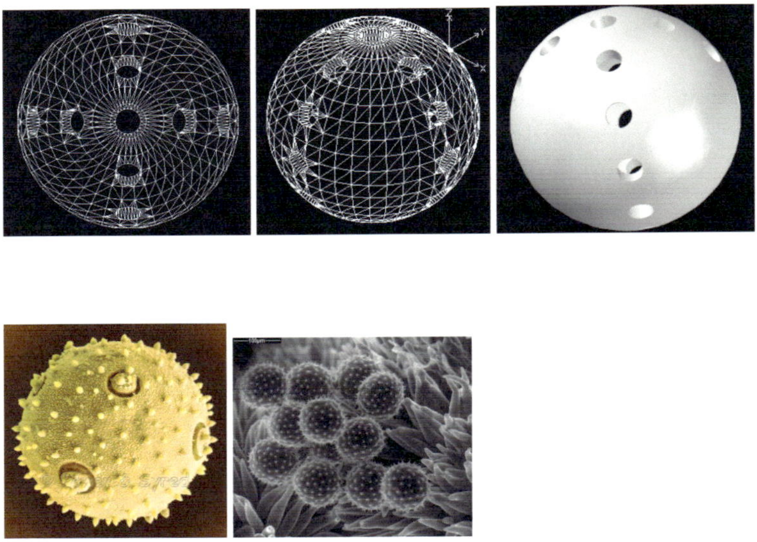

Figure 21 The ball with openings for debris. Scanning electron microscope image of pollen grains from a variety of a morning glory (Ipomoea purpurea) and a cotton pollination.

Figure 22

Figure 23

Figure 22-23: The replicator robot made of small pieces of metal - like technic Lego blocks. The replicator is equipped with a sensor and a micro-computer. On the right: The photo of a spider.

The balls are made of a special metal that does not burn in the atmosphere when falling down. The computer system is calculating the fall of the balls to the chosen coordinates, by using small satellites (Figure 16) to foresee, where the ball must land on the earth.

THE PROJECT - EARTH

This is always a little risk of the wrong landing of the ball. But it is the only risk and danger for the humans, when the ball falls down in New York City instead of in Alaska.

I would like to suggest, that the 3 proposal projects that I wrote about did not had been calculated. The cost of each project is unknown for me. The question is:

1. Who will support the project and what is the maximum donation / financial support?

2. Who takes responsibility for the chosen project?

3. Who takes responsibility for the material loss or man-loss in space?

4. Who wants to take a risk and try to send debris in the balls to the Earth?

5. Who is cleaning after whom, and why?

Space debris is a treasure for many manufacturers. The project pays off in the long term. Many new ideas will come up to the ground and the new projects will start. The technology is moving forward with a support of science. The space debris can be used in recycling to produce metal of higher quality for commercial use or tools for hospitals, the production of high-quality jewelry, furniture, cookware, household machines, appliance, electric appliance, computers, robots and much more. It is because of its high-quality nanoparticles and special manufacture.

LINKS

- http://www.deorbitaldevices.com/site/
- http://www.universetoday.com/89869/nasa-to-test-new-solar-sail-technology/
- http://gomspace.com/
- http://www.dailymail.co.uk/sciencetech/article-1309410/Mission-heart-Sun-Nasa-send-probe-Suns-atmosphere-time.html
- http://vbn.aau.dk/da/clippings/loesning-paa-rumskrot(19e12229-a4cf-49c1-9357-9e38c7fef5a0).html
- http://www.deorbitaldevices.com/site/about-space-debris/
- http://citeseerx.ist.psu.edu/viewdoc/download?doi=10.1.1.452.7457&rep=rep1&type=pdf
- https://en.wikipedia.org/wiki/Space_debris
- http://www.nasa.gov/mission_pages/station/news/orbital_debris.html

Malgorzata Juliet Stach, 23.03.2016

Made in United States
Cleveland, OH
05 November 2026

47646342R00026